How to Make Your Science Project Scientific

Thomas Moorman

AN ALADDIN BOOK
Atheneum

PUBLISHED BY ATHENEUM

PUBLISHED SIMULTANEOUSLY IN CANADA BY
MCCLELLAND & STEWART, LTD.
MANUFACTURED IN THE UNITED STATES OF AMERICA BY
FAIRFIELD GRAPHICS, FAIRFIELD, PENNSYLVANIA
ISBN 0-689-70452-6
FIRST ALADDIN EDITION

How to Make Your Science Project Scientific

To Margaret

Contents

How to Make Your Science Project Scientific

What Does Scientific Mean

Most people care only about the results of science; they care little (or know little) about the methods. Yet without the methods, the results are meaningless. Are the results truthful? We cannot say until we know the methods used to find out those results. Are dogs more intelligent than hogs? Most people say, "Dogs are." But what have been their methods for finding this answer? It's easy to see bias or prejudice here. The dog has been enjoyed as a friend and playmate. Hogs and pigs, if enjoyed at all, were as food; otherwise hogs are thought of as dirty, and above all, stupid animals. Such a prejudice builds up so gradually most people are not aware of having it; they think they know that dogs are more intelligent than hogs.

The *method of observation* is the key to learning the truth about such a question. How do you go about making fair observations? How do you make observations in which your prejudices or biases do not affect the results? Methods are most important. And when you can be reasonably sure that your methods of investigation are going to help guarantee honest results—results not affected by any prejudgment you may have made—then you may say you are being scientific.

Most people have not learned that handling this prejudice problem is important; that prejudice is what makes our ordinary, everyday knowledge unreliable; that superstition and a lot of other untrue "knowledge" is kept alive in a prejudice-system of getting to know things; that control of the prejudice problem is a key difference between ordinary "common sense" knowledge and scientific knowledge.

So, if you want to work on a science project and want it to be really scientific, you must be sharply aware, at every step, of your own prejudice about the outcome. However, you must not believe that prejudice is altogether wrong. You can't help being prejudiced about things. Let's say you like dogs and don't, really, know anything about the intelligence of pigs and hogs. OK, so you have a prejudice; you'd rather like to find that dogs are more intelligent (let's say). How can you test the intelligence of both kinds of animals without letting your prejudice affect the results? Once you plan a satisfactory method for doing that, you are getting scientific about it.

But science is more than just planning how to get an answer to a question. It is more than using your reasoning or your logic. Science means the actual observing of things, of events, of phenomena. You can't be scientific until you actually do some observing of your subject. Your observing, of course, means more than just seeing. It means using all of your senses. It means using measuring instruments such as rulers, meter sticks, balances, clocks, thermometers, etc. It means using instruments that amplify the stimuli your senses receive, such as telescopes, microscopes, sound amplifiers, oscilloscopes, etc. It means using records of many kinds beside pencil or pen—cameras, tape recorders, etc.

This keeping of records is an important part of being scientific because it helps you to *report to others* accurately and honestly about your observations. That helps them to check up on you, which also helps to keep you from letting your prejudices affect your results.

"Objective" is the word for keeping your prejudices or biases under control. In science, being objective means that your findings are not shaped by your feelings, your emotions; that others making similar observations find that they can agree with you even though they may have different feelings, emotions, or prejudices about the results of the investigation. When observers agree on the results in spite of differing prejudices, the results are objective. This, then, is why your records are important and why you must plan to report to others on both your *methods* and your *results.*

Common Sense

A living human being has countless experiences every moment. Some of these experiences are within the person and some are between him and the surrounding environment. A person cannot possibly notice or observe or report all of these experiences, even though the experiences may be changing him in many subtle ways. One cannot notice every heartbeat, every leaf falling, every mosquito buzzing, every drop of rain. And so a person selects or chooses the things he will notice or observe about himself or his environment.

And thus we find that we observe our surroundings with bias or prejudice. What you select to observe another person will completely miss. Or what you fail to observe one time you will closely study another time.

How can we know that the things we choose to know—to learn—are a good, a true, a valid selection of all possible things to know? All of our past experiences come to bear on this choosing, this valuing. Our experiences with other people help to decide our attitudes, our values. And other people help to bring us things to learn from their own experiences and from past generations of people. Some of these kinds of knowledge we may get from others are valuable and some are not. How do we decide about it?

We need a name for the things people claim to know, true or not, by ordinary, widespread experiences. The usual name is "common sense." We must use this term carefully. It has two sharply different meanings: (1) A person who makes careful judgments based on sound information is said to use "common sense." (2) Any ideas or concepts widely accepted by many people but not carefully tested for their truthfulness—such ideas are called "common sense."

It is common sense, for example, to say that heavy things fall faster than light things. Many people accept that as true although a little careful thought should show that it is not a good statement—even without scientific testing. There are many other common sense ideas, some good, no doubt, and some not so good—and all needing testing. Why can't we just accept common sense principles as OK?

The problem is that there are too many variables, usually, along with the bias or prejudice problem. What do we mean by too many variables? Let's work on a rather simple example—the one about heavy things falling faster than light things. People observe "light" things falling —things such as feathers, dead leaves, inflated balloons. They fall rather slowly. And people observe "heavy" things such as rocks, metal tools, chunks of wood. They fall rather fast. So they say, "Heavy things fall faster than light things." But they have not noticed that two or more variables are confounding the results.

A toy balloon, when inflated, is said to be light and, yes,

it falls slowly. If compared with a rock, we must agree that the rock falls faster than the balloon. But we can see two main differences or variables: (1) difference in weight and (2) difference in area or size. We can take a balloon and easily arrange an experiment with one variable only—size (or area). We simply use the one balloon for two trials, first inflated, then deflated, in each trial dropping the balloon from the same height. Now we observe that the balloon inflated (larger area) falls more slowly than the balloon deflated (smaller area); yet the weight is about the same. We might make an even better experiment if we got two balloons, same shape and size, and inflated one and dropped the two of them (one deflated) at the same time from the same height side by side.

Such experiments make us doubt the common sense principle that heavy things fall faster than light things. We see that size (or area) is an important variable.

We might go ahead and plan another experiment with balloons making size or area the same and making weight the variable. To do this, let's say that you would, again, use two balloons the same size and shape. You would make one heavier than the other by putting into it a weight—perhaps a penny. Then you would inflate the two to as nearly the same size as possible—making area constant. You would drop the two from the same height, taking care to release them at precisely the same time. In this case, we see that the heavy thing does fall faster. So we have one experiment that does not support the common sense principle and one that supports it.

A third experiment is needed. We would get a piece of metal that was lighter than the experimental balloon—a piece of wire or such. Then we would drop the inflated balloon and the piece of wire at the same time and observe the results. In this case, the lighter thing (the metal) would fall faster than the heavier thing (the balloon). So by three rather simple experiments we would have shown all of the following:

A. Two things of the same weight falling at different speeds
B. A heavier thing falling faster than a light thing
C. A lighter thing falling faster than a heavy thing

With this conflicting evidence, it appears that something is wrong with the common sense statement "Heavy things fall faster than light things."

The statement is too simple to cover the differing results of our three experiments. One might go further and use heavier objects than balloons (as Galileo is thought to have done). An example would be to compare the falling speeds of wooden balls of the same size and shape but with one hollowed out and the hollow filled with lead. And one might use more complicated equipment to create a space without air (a vacuum space) and arrange for things to be dropped there to compare their falling speeds. With enough good experimenting and observing, one might then make a statement that would correctly express the falling speeds of things light and heavy, large and small and in air and in vacuum. We see, however, that with many variables all present at once as in ordinary observation, the results are confounded—are mixed up too much for an accurate, truthful statement to be made.

Not only are there too many variables for truthful results in much common observation but there often is too much chanciness. For example, common explanations for catching a cold are, "I got my feet wet" or "I sat in a draft and got chilled." These have been shown by careful testing to be unreliable as causes of colds. However, a person may have got his feet wet or got chilled by a draft at just the "right" time for it to seem that one of these caused a cold. It is this sort of chance happening of things at certain times and places that often seems to confirm a common sense principle even though other and better testing may show the principle to be wrong.

It will not do, however, to say that all common sense principles are wrong. Many have a "grain of truth." It may not be clear, however, which are the truthful parts and which are not until someone performs tests good enough to be called scientific.

Before we go further in our search for the meaning of "scientific" we should consider a kind of evidence often called *anecdotal evidence.* Sometimes a person brings to a discussion, as evidence to support a proposition, a story out of his personal experience. Such a story (or anecdote) may seem very convincing to the person telling the story and to his listeners. Nevertheless, as evidence, it may have all the weaknesses of common sense. The events in the anecdote may have happened suddenly, without plan. And the person telling about the events may have had only a narrow, prejudiced view of the happening.

But suppose the person is honest and accurate in giving his anecdotal evidence. Suppose, let's say, that he did take vitamin *Z* all winter and did not have a single cold, while people all around had the usual colds. Are we going to accept this as evidence that vitamin *Z* prevents colds? Not likely. There is the chance that he might have gone through the cold season without colds anyway, even if he had not taken the vitamin. It takes a broader and more thorough investigation of this sort of thing before we can conclude that we have established a good general principle. And so, we must be wary about accepting anecdotal evidence because of its chanciness as well as because of the possible prejudiced nature of the observations.

There is a place, of course, for observation of people and things under natural conditions. But one must plan on how to make the observations objective and the results reliable. We will consider these methods further in the chapter "More Scientific Methods."

Simple, Uncontrolled Experiments

In science there are special methods for observing known as experimental methods or "doing an experiment." Most people think science *is* experimenting. Not so. However, since this is a most popular scientific method, let's work on it first. Let's see what it takes to make experimenting scientific. First we will analyze simple, uncontrolled experiments. Then we will go on to more complicated kinds designed to help get good results under more difficult conditions.

To help you do a good job with experimenting, we need to analyze or dissect an experiment and find names for the parts.

First, of course, you are the *experimenter*. In doing an experiment you change something to observe what happens. There are two important changes in an experiment:

1. The change you make as an experimenter.
2. The "what happens"—the change if any that follows the change you made as experimenter.

The change you build into the experiment is the *independent variable.* The "what happens" that follows after is the *dependent variable.* As an example, let's say you happen to be sitting watching a candle burn. You ask yourself, "What would happen to the color of the flame if I put green food coloring in the pool of melted wax? Would it turn the flame green?" If you were to go ahead and do the experiment, the independent variable would be your putting the food coloring in the melted wax. The dependent variable would be the change (if any) in the color of the flame. The color of the flame depends on what you do to the wax, not the other way around.

If the flame does change color when you add the food coloring to the wax, we say that the independent variable and the dependent variable are *related variables.*

As experimenter you must try not to let other things change beside your two main variables. Let's say that you want to learn if adding baking soda to the water used for watering a bean plant will make it grow better. You must decide on how much baking soda to put in the water and carefully observe the growth of the bean plant before and after. Obviously, you must not make any other changes such as adding more water or using a different temperature or a different amount of light. Other variables such as these would *confound* the results; you would not be able to judge which of the four variables—baking soda or more water or different light or different temperature—might be related to any difference in growth. So you try to keep other possible variables *constant* or the same during the experiment. You try to keep all possible variables constant except the one you built into the experiment—adding the baking soda to the water as your independent variable.

Now we come to the problem of prejudice again. Most scientists do an investigation with a hope that it will turn out one way, not another; that it will "prove" something they'd like to show to others. This is a bias or prejudice in favor of one result instead of another. Scientists have found that it is good practice to state this hoped-for result ahead of time as a *hypothesis.*

The important thing about this is to freely state the hypothesis and then to try to plan the investigation in such a way that prejudice does not influence the outcome. Making the investigation proof against the prejudice of the investigator is not easy to do, especially in the more complicated kinds of investigations such as those in biology or in human behavior. So, scientists have worked hard to develop more reliable kinds of experiments and better methods for analyzing the data, the results. A *controlled experiment* is just one step in this direction.

Controlled Experiments

Changing from a simple, uncontrolled experiment to a controlled experiment can give your findings a great boost in believability. The word *control* has a special meaning here. Of course the investigator controls most of what happens in his experiment—or he hopes that he does. But there is a further meaning.

We discussed adding baking soda to the water used with a bean plant to see if the plant would grow better. Many things could go wrong with only one bean plant, so it would be a better experiment if you used, say, ten plants in a larger planter. But even then something could happen to the planter that you would not know about that would confuse the results by changing the growth pattern of the plants. So, to make a better experiment, you would use two planters, each with ten plants. You would arrange all things about the two planters to be as constant as possible, except the independent variable—baking soda in the water for one planter. Constants for the two planters would be:

the kind of seed, the kind of soil, the size and shape of the containers, the amount of water added each time, the temperature, light, etc.

Now you would have a controlled experiment. One planter—the one to which you added the baking soda in the water—would be called the *experimental planter;* the other would be called the *control.* Two planters like this take more time and materials than making just one. There are important gains, however. If there were some influence at work (unknown to you, the experimenter) that changed the growth pattern of the plants, it would probably happen to both groups and you would know it was not the baking soda. Further, having the two planters side by side helps to make comparing the two more accurate and easier; it helps you, as experimenter, to judge more truly the difference (if any) that the baking soda in the water makes.

Suppose you did not use two planters but first planted ten bean seeds, watched them grow, kept accurate records of the water, temperature and other constants. After the plants matured, let's say, you then rooted them out, replanted the planter with ten more seeds and, this time, used baking soda in the water as your independent variable. Again you kept accurate records to compare with the first planting and, in the end, you might be able to decide that the baking soda made an important difference. This kind of experimenting is what most gardeners and farmers do—an experiment that might be called a controlled experiment but in a before-and-after design instead of a side-by-side design. Unfortunately, in the before-and-after design, unwanted variables may creep in, especially in farming, when the experimental planting is done a year or more after the control planting with which it is to be compared. And so scientists think of a controlled experiment, usually, as one where the two groups are together in time.

In setting up a controlled experiment, many decisions must be made. Take our example of the experiment with beans. Let's say that you decide to use ten beans in each planter. Which seeds will you use in the experimental and which in the control? You will take all of the seeds from the same package, of course. Of course? There, you see, is an important decision. You want the seeds to be as much alike as you can get them—as alike in inherited traits as possible. If you grow your own, you might take them from the same plant. But it is more likely that you would buy a package of seed beans. So you take the 20 from one package of seeds. As you pick them out, you are careful to get uniform, sound beans (no broken ones, undersized, discolored, etc.).

Now even here it is possible for prejudice to begin to affect results, so you do not decide yet which planter will be experimental and which control. You will use a *random choice* method in making this decision, as in making others to come later. And in choosing the seeds, you put the seeds—the 20 you have selected—into a separate container where you cannot see them. Then you take them out random choice, "blind," as in taking names out of a hat, and you put one for this planter, one for that.

At this point some people might say, "This is ridiculous! Why all this bother? Why not just plant the seeds and get on with it?" Of course, if one were just planting seeds for the fun of it or for growing food, one would not worry about doing it by random choice, in an unprejudiced manner. We must look ahead, however, to a final judgment that will depend very much on the making of random decisions at this early stage. And when you report your findings, you want to have full confidence in your methods.

You continue, therefore, (using our bean experiment example) to make as many decisions as you can randomly. The soil you would use in each planter must be mixed thoroughly in one large container and then transferred, some to this and some to that planter, alternately in small amounts so as to not favor either. You plant the seeds spreading them in as near the same pattern around each planter as you can, and you take care to cover them uniformly with soil. All is ready.

Now you may choose which is to be control (no baking soda in the water) and which experimental (baking soda in the water). To make an unprejudiced decision—a random choice—you flip a coin; heads, this is experimental, tails that.

The rule, you see, is to arrange to make as many decisions randomly as you can, tossing a coin, pulling numbers or names out of a hat, rolling dice, using a list of random digits, or other means. When you make decisions randomly, you are saying, first, that even though you held constant as many variables as you could, yet you realize that there are probably other unknown variables that you could not make constant. These might be such things as differences in the genetic traits carried by the seeds, differences in the soil about which you may not know and so on. Therefore you are agreeing that you will take your chances on these unknown variables coming out for or against your hoped-for result—your hypothesis.

Can we ever make the conditions exactly alike in the two parts of a controlled experiment—aside from the independent variable? Not likely. Nevertheless; we can count on a controlled experiment to give more valid results than a simple, uncontrolled experiment would give or the usual mixed-up conditions of ordinary, everyday, experience.

Counterbalancing Controlled Experiments

You want to settle an argument among friends about the question: Does loss of sleep reduce a person's physical strength? You decide to test the hypothesis: Sleeping only four hours in a night instead of eight will increase the time it takes a person to run 100 meters the day following. There are many problems about this experiment that we will not go into now; let's just keep it simple for an example of *counterbalancing*.

You have eight friends who agree to work with you. For the first night, you divide them into two groups. Group 1: All four will sleep eight hours; they will go to bed at 10:30 P.M. and get up at 6:30 A.M. Group 2: All four will sleep only four hours; they will go to bed at 2:30 A.M. and get up at 6:30 A.M. At 6:30 A.M. you will take all of them out to the athletic field and, using a stopwatch, time each on a 100-meter run and record the time.

Now for the counterbalancing (also called cross-over design): On the second night, the Group 1 subjects (who got eight hours sleep the night before) sleep only four hours. The Group 2 subjects sleep eight hours. (See Figure 5.1)

Figure 5.1
Counterbalancing Plan

Trials	Control Group (*8 Hours of Sleep*)	Experimental Group (*4 Hours of Sleep*)
Trial 1 (1st day)	Group 1	Group 2
Trial 2 (2nd day)	Group 2	Group 1

When all the running is over, you will average the time for all eight runs for all eight-hour sleepers and for all eight runs for all four-hour sleepers and try to determine if there is a meaningful difference. The counterbalancing is used to take care of differences that may already exist in the two groups. It would be practically impossible to find two groups so nearly matched in running ability and other characteristics as to make a fair test using only a controlled experiment, without counterbalancing.

Later as you study other aspects of judging the value of experimental results, you will be able to point out serious problems in this experiment, as outlined here. But the example does show you how the counterbalancing principle operates.

A simple experiment (uncontrolled, not counterbalanced) may be good enough for testing a hypothesis about fairly simple physical things. But for experiments in biology and in the ways people and animals behave it is difficult to pin down constants so that only the variables we want are varying. So we need better experimental plans —controlled experiments with counterbalancing.

Will this pet mouse eat this dried pea? A simple, uncontrolled experiment will serve. He eats it or he doesn't.

Can mice live healthfully on dried peas? You would want a controlled experiment: a control group fed a good normal diet and an experimental group (as evenly matched as possible with the control group) eating only dried peas.

Is there more fighting among mice living in crowded conditions than in less crowded? Here you might better use a counterbalanced plan. You would have carefully matched groups in two cages or pens, one cage or pen larger than the other but with everything else constant— food, water, litter, ages, sex, light, temperature. Then after a few days of observation and recording of "fighting" in each group you would switch each group to the other cage and observe again for an equal time.

As you grow more knowledgeable about this sort of thing and as you learn more about how professionals do experiments, you will see many ways to improve your own. One of the experiments described earlier is especially open to criticism—the experiment to test the effects of loss of sleep on running speed. You, the experimenter, are aware of the prejudice problem but how about the subjects of the experiment—those doing the sleeping and running? Do they have prejudices about the hypothesis? Undoubtedly some of them would have. And, if so, how would their prejudices affect the results? Some of them might run harder after the short sleep just to show that they were "tough" (were not affected, they think, by a little loss of sleep). Others might be prejudiced the opposite way. This is a difficult problem. It is sometimes handled in the experimental designs called "blind" and "double-blind."

Blind and Double-Blind Experiments

Sometimes experimenters have a problem in experimenting with people: the people know too much. Then it becomes necessary to plan so that the people who are a part of the experiment are kept in the dark about the independent variable.

Suppose you are investigating the question: Do people do mathematics problems better after drinking coffee or not as well? Here you may well have the problem of prejudice on the part of subjects in the experiment. One subject may already believe that drinking coffee helps his math output; another subject may believe that drinking coffee lowers his output. How do you design an experiment to take care of such prejudice in the subjects? The following description of a plan is oversimplified for clarity. Just remember that there would need to be more subjects than there are described here, along with other changes, in order to make the plan good enough for a real experiment.

Pretend that you are the experimenter. You choose two

friends who will cooperate. Both subjects are coffee drinkers and both are pretty good at mathematics. You plan a counterbalanced experiment because you cannot be sure that the two are equal in math ability. They agree to go without coffee or other caffeine-containing drinks or pills for forty-eight hours before each test. You (the experimenter) work out other details such as the number and kinds of math problems and difficulty, the time you will allow, the constants of the test conditions, and how you will supply the coffee. It is the caffeine in the coffee that is the independent variable. Now you decide to use a plan known as a blind experiment. You do not want the subjects to know when they are getting coffee with caffeine in it and when they are getting coffee without caffeine. You use the design of Figure 6.1.

Figure 6.1
Counterbalancing Coffee and Math Experiment

Trial	Subject Getting Coffee with Caffeine	Subject Getting Coffee without Caffeine
First Day	Bill	Joe
Second Day	Joe	Bill

Each subject gets to drink coffee before he takes the math test, but each is "blind" as to the independent variable, the caffeine in the coffee. On the first day Bill gets coffee with caffeine. Joe does not. You give them the test. On the second day, Joe gets caffeine-coffee and Bill gets coffee without caffeine, and again you test them.

As you do the experiment, you hope that you can keep Bill and Joe ignorant as to which of them is getting caffeine and which is not. However, a person may be especially aware of his own feelings and other reactions when getting caffeine so this may not be a reliable blind experiment. Nevertheless, it shows the pattern. There is no problem in this experiment with the experimenter's prejudice if he uses an objective mathematics test to determine the results. An objective test means that there is no chance for bias or prejudice in scoring the test. That is, the test is prepared ahead of time, all the problems are clearly stated and the solution for each can be clearly decided right or wrong from a scoring key. A nonobjective test would be one in which, for example, people were asked to write humorous stories and the tester would try to decide which were more, and which were less, humorous. The prejudices of the tester, or experimenter, could then become the main element in classifying the stories as better or not so good. His decisions are more subjective, you see, than decisions made by some kind of measuring plan or by the scoring key in the math test.

In some experiments the experimenter must be "blind" as well as the subjects. This arrangement would be called a *double-blind* design. The need for this occurs more often in medical research or in some kinds of psychological experiments than in ordinary physical or biological work.

Suppose, for example, you were a medical researcher with a plan for investigating this question: Does ingredient *Q* when added to suntan oil protect the skin from sunburn more than ingredient *K*?

You would, of course, arrange for a large number of subjects to try the two preparations under your supervision. Each would be given a bottle of the experimental suntan lotion. No subject would know which of the two experimental ingredients was in his particular bottle. Now, your problem as experimenter is to judge the amount of protection from sunburn for each subject. Your *judgment* is critical in the honesty of the rating received for each ingredient. You plan, then, to not let yourself know which subject gets which ingredient in his suntan lotion until you have done all the judging at the end of the experiment.

To make your part of it blind, as experimenter, you have a trusted friend prepare the bottles of lotion. He identifies each by a code number so that only he can tell which ingredient, *Q* or *K,* is in each bottle. You, the experimenter, do not know. He puts this record away for safekeeping until the experiment is completed.

Then, you, the experimenter, go ahead testing all the subjects and making a record of the amount of sunburn each experiences, of course making sure that each uses the bottle of suntan lotion assigned to him. You put the bottle number on the record with your judgment of his sunburn.

After it is all over, your friend gives you the secret, coded record. You then divide the records of the subjects into groups—those who used ingredient *Q* in one group, and those who used ingredient *K* in the other. You had not known which individuals belonged to which group until you identified which ingredient each had used. Now that you have the groups identified, you compare the degree of sunburn that you had previously judged "blind" in each group (your dependent variable) and perhaps learn if there is a significant difference between the two groups.

Suppose, however, that you had wanted to test only one

ingredient for its effect in preventing sunburn. Let's say it is ingredient *Q*. How do you make such an experiment blind (or double-blind)? You would need two suntan lotions and two groups of subjects, of course, but one of the lotions would have no anti-sunburn ingredient. This would be called the *placebo*. The two lotions would have to be made to look, feel, smell, and taste so much alike that the subjects could not tell the difference. Then you would proceed with the experiment as before. Recall the plan for the testing of coffee (or caffeine) and doing math problems; the coffee without the caffeine was a placebo, too. It is especially important to apply this method in the experimental testing of medicines, vitamins and the like. Why is this important? It is the prejudice problem again. People sometimes have such strong belief (prejudice) about a medicine or a vitamin that even when they only think they are getting it, they report that they are getting the results expected. Even though they are only getting a placebo—a "look-alike" which has none of the medical or nutritional power of the real thing—they show a response or a recovery from their symptoms. This happens enough times with enough people that scientists working in these fields insist upon blind and double-blind experimental plans to help make the results more valid.

This also warns us that we must be careful about accepting anecdotal evidence that people report from the use of medicines or from special diets, vitamins, exercises, etc. Such self-experimenting is not necessarily wrong. It is just that we cannot be sure that the results are related to the particular variable that the self-experimenter claims. There are too many variables not accounted for, no control (in the controlled experiment sense) and always the prejudice problem.

More Scientific Methods

Science involves experiments but it is more than experimenting—much more. Some people think that experimental science is the best "science." That is, they believe that other methods do not prove things as well as experimenting does. Or, they just like experimenting as a way of investigating more than they like other methods.

But your project is, let's say, astronomy. You cannot experiment with the stars! You cannot even be sure that the star you see is "there." The light you see may have been on its way here for hundreds of years. Where is the star now? All you can do is to record and report your observations. Of course the kinds of equipment you use for receiving and analyzing the light or other radiation from the stars offer opportunity for much experimenting. Nevertheless, in the end, all you can do is observe the star as it exists (or did exist). You cannot change it to see what happens.

This observation of things as they exist in nature is

naturalistic observation. Bird watching and other "nature study" is largely naturalistic observation. Much of our knowledge of plants and animals has been gained by observation of things in their natural ways of living. We should not try to decide in any broad general way which is best, naturalistic observation or experimenting. They are different methods and you should try to choose the method that best serves your purpose in the investigation.

Do you want to find out how field mice of a particular species live through the winter? Or do you want to test the intelligence or learning ability of the mice as compared with other animals? Naturalistic observation is best for one investigation, whereas laboratory experimentation would be best for the other.

Naturalistic observation is difficult to do "scientifically." We all grow up using naturalistic observation most of the time to learn about the world around us. Many of the wrong things we learn, including our prejudices or biases that stand in the way of our seeing things as they really are, come from naturalistic observation. How does the observer keep his prejudices from affecting his conclusions in naturalistic observation?

One method of scientific observation is called *case study.* The medical doctor's observation and treatment of patients is an example. Case study was especially important before modern experimental medicine developed. A doctor's cases are a blend of naturalistic observation and experimenting. When a patient comes along with a particular problem, the doctor "treats" the patient to see what happens. This is an experiment, although a doctor usually avoids calling it an experiment in the presence of a patient. In his usual office work a doctor cannot easily set up controlled or counterbalanced experiments. He must depend, instead, on his experience with other cases he has handled or on reports read in medical journals or books of other cases and experiments. Since his primary goal is to help a particular patient, he must put into the background any goal he may have of scientific advancement of medicine.

In the study of human behavior in psychology, sociology and anthropology there are limits upon experimentation. We cannot experiment with people in such drastic and perhaps destructive ways as with physical objects or with plants or some animals. So, for much of the study of people we are limited to naturalistic observation, case study, and surveys.

The term *scientific rigor,* or just *rigor,* is used in discussing the various methods of science to mean that some methods may permit more strict, more severe or more sensitive tests of the hypotheses than other methods. Also, of course, one investigator may be more rigorous in his work than another. That is, he is able to make his methods more scientifically objective. But how does one go about making naturalistic observation, case study and surveys scientifically rigorous?

Suppose that you are deeply interested in the subject of your observation—a person, an animal, rocks, insects, photographs of stars or whatever. You are excited; you have discovered something that thrills you; you feel joyful, poetic. Very good! A scientist must admit being moved by feelings, by emotions, just as anyone else is. Nevertheless, our feelings help to trap us into reporting things with biases for (or against) certain viewpoints.

So, while you are doing your observing, you try hard to be rigorous, to be objective, to try to record your observations with as little bias as you can. And then you prepare a report of the results. In this you try to tone down the "feeling" content. You try to emphasize the objectively observable things—the things about which any other qualified observer would want to know so that he might check up on you in a similar situation.

A step beyond naturalistic observation are two more important science activities, *collecting* and *classifying*. Though collecting specimens is a step away from naturalistic observation, the two activities often go hand-in-hand, especially when it comes to the collection and classification of rock and mineral specimens or plant or animal specimens. And a person who has made several observations of anything has a collection of observations, even though he may not have gathered specimens.

Scientific classifications use names or numbers or other symbols to organize like and unlike materials. Finding names, or making up names, becomes an important problem in some fields of science, in chemistry or biology, for example. But, of course, the names do not come first; the first problem is how to usefully classify the things.

Think about a collection of several dozen, or several hundred, rock and mineral specimens. There is no single, correct way to classify them, even though there already exists a scientific classification that may appear to be the "correct" one. A collection may be classified in many different ways according to any one or more of the following characteristics or criteria:

—size of specimen
—color
—crystalline form
—chemical composition

—hardness

—density or specific gravity

—fossil content

—geological origin

—geographical origin

—money value

— other criteria

The needs or purposes of the person doing the classifying must determine the kind of classification used. This does not mean that there is something wrong with a standard scientific classification. Scientists, too, have important needs and so their classifications usually help to satisfy those needs:

—to help to show relations between different subgroups (as species, genus, etc. in biology)

—to help to make communication about the things clear and accurate; and so the classifications are accurately described and special names used.

There are good reasons for using scientific names for such things as flowering plants, for example, because of the confusion of common names. And since scientific classifications and names are widely published, many people find them useful for nonscientific purposes. You should know about them and use them where they will do. But, even so, you may find that you have very good reason for working out your own classification for your own special purposes. For example, you might classify insects in your locality several different ways:

—those that attack garden plants and those that do not, or are beneficial;

—those that are "pests" to campers, picnickers, etc. and those that are not;

—those that attack other insects and perhaps help to control them, and so on.

Is such a collection and display a science investigation? Not necessarily. It may be only a reporting of facts or information that other people have found out.

What does it take to make a collection (or collection and classification) acceptable as a science investigation, firsthand? Collecting things to test a hypothesis or to "prove something" may also be called a survey. The next chapter will tell you how the survey may be used as an important scientific method.

The Survey as Scientific Method

A simple survey would be to count the number of people living in an area (a few blocks, a small town). Less simple would be a survey that tallied and analyzed the inhabitants by age, sex, size of household, vocation, and the like. Or one could take a count of the different kinds of trees in a wooded area or of the bird and other animal species, and that too would be a survey. Most people would not think of these as science projects. However, they could be done poorly and inaccurately, or well (more completely and accurately), and to the extent that an effort was made to do them well, they would be called scientific. These would be called censuses or enumerations and would count as naturalistic observation.

A scientist is usually trying to test a hypothesis, trying to test its validity. In this sense, a survey may be like an experiment. Of course, in an experiment, the experimenter changes something to observe what happens. Most surveys, however, are different from an experiment in this way:

the scientist tries not to change the things he observes. The things have already happened and as he observes them he takes care to disturb them as little as possible by his method of observation. In a public opinion poll or survey, for example, if the survey is to be scientific, the scientist plans his questions as well as he can to find out what the people already believe; he tries to avoid asking questions in ways that might change their beliefs.

Some surveys, however, may be combined with an experiment; a survey may be made of opinions before a publicity or propaganda campaign and another survey made afterward in order to learn the effects of the campaign on opinions.

A survey may include related variables, an independent and a dependent variable, or other relations. Think of an investigation based on this question: How much is the playing of musical instruments by children related to the playing of instruments by their parents? Consider how this would have to be done as an experiment. We would have to rear children in a prepared plan—some with parents who play musical instruments and some with parents who do not play. After ten or twelve years we would compare the groups to see what we could learn. No one, obviously, is going to do such an experiment. The only workable method is to take a survey of many parents and children. As in an experiment, the hypothesis would include a statement of two variables that the investigator suspects to be related: (1) playing of musical instruments by parents and (2) playing of musical instruments by their children. The hypothesis might be: "There is a statistically significant relation between the playing of musical instruments by children and the playing of musical instruments by the parents of the children."

As in an experiment, a survey may require that constants be maintained. Thus, in our example, the children would in the end be classified into groups: those who play instruments and those who do not. However, we would not want to allow biases to creep into the findings. Therefore we would try to see, for example, that all of the children came from families of similar income level, similar educational level, etc. Or, as a different approach, we would try to do a good job of randomizing the families so that all income levels, educational levels and the like would be fairly represented among both groups of children.

To get information for such a survey would require interviewing the children or their parents face-to-face, by telephone, by mail or in some manner with a suitable questionnaire. In the end, we would hope to have enough information on enough children and their parents to make a decision about the hypothesis. Would an affirmative answer ("Yes") to the hypothesis tell us that the parents' playing was the cause, or the main cause, of the children's playing? Probably not. The causes of such playing are complicated. There are many other variables that may be part of the answer. Nevertheless, we might find by such a survey that the two variables are related.

People make surveys informally all the time as a part of living. Do girls get interested in boys at earlier ages than boys get interested in girls? Are children from large families better able to compete at school and elsewhere than those from small families? Are men better drivers of automobiles than women?

Often we find people who believe that they have found answers to questions such as these from "experience," that is, from naturalistic observation or informal surveys. However, there is the prejudice problem here just as there is in experimenting. Too often the person accepts his hypothesis, his belief, first and "looks for" cases that support his belief and "doesn't notice" cases that go against it. He says, "I know men are better drivers than women; I see it all the time."

Or he finds his answer by a sort of logic: "Of course a child from a large family can compete better. He has all those brothers and sisters to contend with right from the start. He's not spoiled like an only child. He's got to learn to compete." He uses what seems like logic and thinks he has tested his view by experience.

Such a person is often prejudiced against scientific methods of answering his questions. If someone shows him a scientific study of, say, automobile driving characteristics of men and women, he pushes it aside. He doesn't want to take time to see how the variables are defined; he doesn't sense the importance of large and representative sampling of people in such a survey; he is not prepared to analyze the data in a more valid way. He may say, "Oh, that's too complicated; too much statistics; what do those people know about real life anyway?"

Such a person has not faced the problem of stating his question clearly. What does it mean to say that men are better drivers than women? Can the men drive faster? Or win more races? Or handle more difficult problems such as backing a semitrailer? Or drive longer without tiring? Or have fewer accidents? Or break traffic laws less often?

Men actually have more accidents than women in the driving of cars, trucks and buses. But more men drive than women do; they drive more vehicles, for longer periods and for greater distances. So, how would you make a fair comparison in a survey? It takes statistical treatment of the data. One must figure averages such as accidents per million miles driven by men and women. Just because people, in our culture, encourage men to drive racing cars and discourage women from doing so does not prove that men are necessarily better as race drivers. In some ways women are healthier and have more endurance and are superior in some skills. Perhaps women, if given equal training and experience and equal opportunity to win the rewards, would make as good or better racing drivers than men. Who knows? No one.

It should be obvious that there are many uncontrolled variables that make it impossible to answer some questions —or that statistical treatment of large amounts of data must be used to answer some questions. Do not let this discourage you, however, if you would like to do a survey as a science investigation. There are many that can be done in a scientifically satisfactory method with reasonable time and effort.

Do boys or girls spend more time watching television? Which programs? For a limited population, you could do a valid survey of this in a school or class. You would need to develop a good questionnaire and the methods for interviewing a fair sample of the group—or the whole group if not too large.

Are children more superstitious than adults? This might present a more difficult problem in preparing a questionnaire that would give a valid answer. How about astrology? What fraction of the people believe that astrologers make valid predictions about human affairs? Do more males than females believe in astrology?

In all such questions you are faced with the problem of defining the variables well enough to allow you to get meaningful answers. What do you mean by "watching television"? by "being superstitious"? or "belief in astrology"?

How can you measure things like these? One person may act superstitious by doing such things as "knocking on wood" in a playful mood and yet not be seriously superstitious. Another may be strongly superstitious but not display it so openly. How would you go about framing questions that would bring you valid answers? This is the problem of the survey taker.

After finishing the data-gathering in a survey, there can be a huge amount of work to be done analyzing and tabulating the data from the questionnaires. Consider this when you plan a survey.

After the data is tabulated and available for analysis, then what? Then there is the same problem of evaluating the results as there is in experimental work. Is there, for example, a difference between the amount of television watching done by girls as compared to that done by boys? How much of a difference must you have for it to be significant—to count as a difference? This is something you will have to decide, but the chapter "Evaluating Results" may be of help.

In the discussion of collections of rocks, of insects, and the like, it was suggested that such collections are similar to surveys. A collection may be used to test a hypothesis just as an opinion survey may be used. As an example, suppose two parts of a countryside are separated by a valley. Do they, nevertheless, contain the same "kind" of land, in geologic terms? A collection of specimens of soil and rock from each area when compared might demonstrate the relationship.

Surveys and collections then may make it possible for us to obtain scientific answers to hypotheses that cannot be tested by experiment.

Measurement

For this discussion of measurement, the metric system will be used in all examples. If you have a meter stick at hand for reference it will be useful. If you do not have one, the commonly available foot rules with a metric scale along one side (about 30 centimeters, each divided into tenths or millimeters) will do.

Let's say that you want to measure across a room with a meter stick. You agree to measure it to the nearest millimeter (0.001 meter). If you are not familiar with measuring in the metric system, notice that a millimeter is as small a unit as 1/32 inch (roughly). Now, to measure across a medium-sized room to this degree of accuracy is about like measuring a glass ball—a marble—of 1 inch diameter with a micrometer to an accuracy of about 0.0002 inch (2/10,000 inch).

Let's say that the room you are measuring has a hard floor, which lets you make a pencil mark at the end of the meter stick each time you lay it out across the room. (Or, if there is a carpet, you can stick a pin in to mark the end of the stick.) You find that there are 5 whole meters and a part that looks like Figure 9.1 at the arrow point.

Figure 9.1

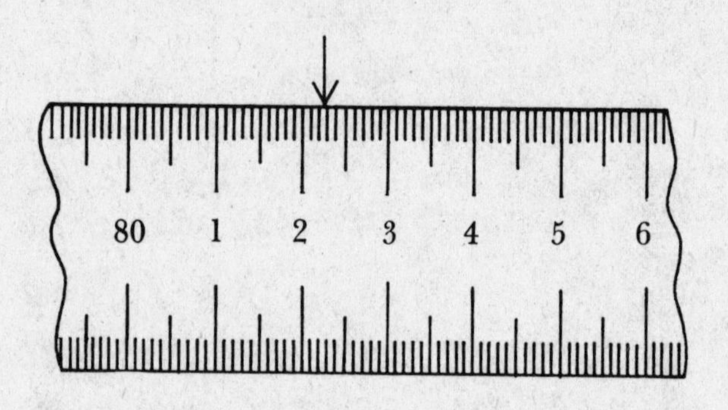

The arrow shows 0.823 meter

This would read 5.823 meters. But one measurement is not enough if we intend to understand the problems of measurement. We should try to get an independent second measurement. To do this, you might have a friend make a measurement without knowing, ahead of time, what result you got.

Suppose you now have two measurements: 5.823 and 5.834 meters. Which is right? No one knows. How about measuring it again? Good, but would that prove to be the correct measurement, any more than the first two? Not likely. No one can know the "true" measurement. We must face the proposition that there is no perfect way to measure a thing. No two people can measure a thing, even the width of a room, exactly alike. Even the same person is not likely to get the same measurement twice, especially if he waits awhile between measurements to allow forgetting of the first measurement.

And, just as important, no two meter sticks or other measuring instruments are exactly alike. There is no such thing as a perfect measuring instrument, just as there is no perfect measurement.

To get as good a measurement as is reasonably possible of the width of the room, it is necessary to take several independent measurements. Let's say you end with five different values: (1) 5.823 (2) 5.834 (3) 5.829 (4) 5.830 (5) 5.825. Now we have a statistical question. What are we going to pick as the width of the room? If we take the arithmetic average (or "mean") we get 5.828 meters. This we see is not any one of the values we got by actual measurement. Is it the "correct" measurement? All we can say is that it is probably very close.

If you were measuring the diameter of a glass ball (a "marble") with a micrometer, you would find the same problems. Several independent measurements would probably give several different values. Again, no one would know which was the correct diameter. And again, there would be a statistical problem in choosing a number to represent the diameter—the "true" diameter—which no one can know.

Scientists have understood this uncertainty about measuring for a long time. It seems to surprise others, however, who have not worked so hard to learn how to make good measurements. Some people react to this discovery by saying, "Well, if I can't really know how long it is, why bother? Why bother trying to make it exact?" However, scientists and others who keep working out better ways for measuring things are not doing it just for the fun of it. They are trying to communicate better. Reporting scientific findings to others is an important part of scientific methods, as you know. And communication among scientists has been helped (immeasurably?) by better and better systems of measurement.

The current system, officially known as SI but commonly called the metric system, has been developing since the French started it in about 1800. Scientists and other experts get together in international conferences to discuss new methods of measuring and to agree upon new standards or to adopt new units for dimensions that have not been standardized before.

Our discussion here has been about measuring the length of an object, only. However, all that we have said about measuring length applies to the measurement of other dimensions or physical quantities. SI (initials from the French, Système International) recognizes these six basic units: Length (meter), Mass (kilogram), Time (second), Electric current (ampere), Temperature (kelvin) and Luminous Intensity (candela). From these come many "derived units." An example is the measurement of area, for which the unit, the square meter, is derived from the basic length unit, the meter. Or we derive velocity (commonly called speed) from length and time measurements as meters per second.

Yet in spite of all this work to develop a more complete system of measurement, and the fact that scientists are steadily achieving more refined standards, we have an uncertainty in all measurements.

How do we deal with this uncertainty (besides pretending it doesn't exist)? In most ordinary measuring of things, one reasonably careful measurement is all that is needed. Most people will settle for that, believing that they know the length of a thing, or the time, or the weight, etc. However, many scientific and technical workers must work as near to the limits of accuracy of their instruments and methods as they can. Then they must report to others, in some way, the limits of the methods they have used. And so there is need for ways to express such limits.

Here is one method: Let's say that you are measuring across a piece of paper with a scale (a ruler) divided into centimeters and tenths. You want to express your measurement as centimeters to the nearest tenth such as 23.7 cm. (See Figure 9.2).

Figure 9.2

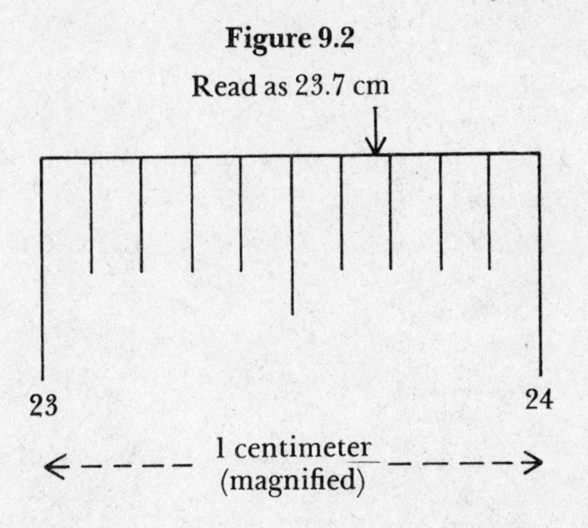

As you measure, you see that the edge of the paper is not quite on one of the marks showing tenths so you pick the tenth mark that the edge is nearest to. That is, you pick 23.7, not 23.6 as in Figure 9.3.

Figure 9.3

Read as 23.6 cm

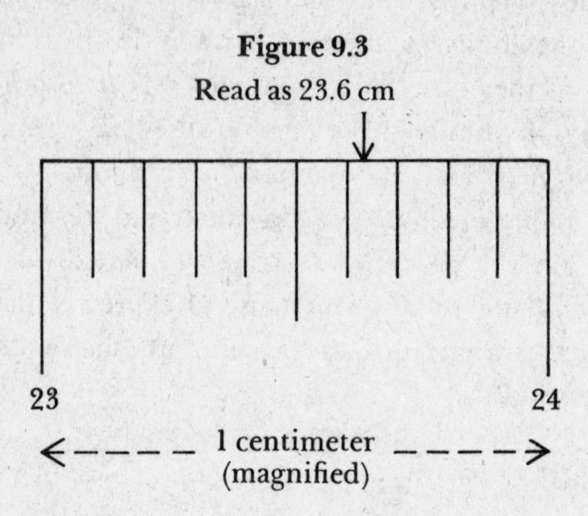

However, you do not want anyone to think that you are claiming that the paper measured exactly 23.7 cm. What you would like to tell others is that you are reading the scale to within ½ of a division either way from a scale mark; (See Figure 9.4).

Figure 9.4

Read as 23.7 cm

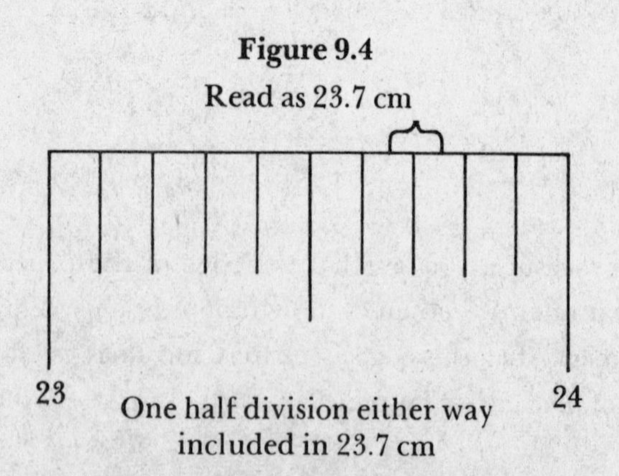

Now ½ of one-tenth is 0.05 in decimal notation. So you put ± 0.05 cm after your measurement figure thus: 23.7 ± 0.05 cm (Read, "23.7 plus or minus five hundredth centimeters").

For another method of expressing uncertainty, go back to the measurement of the width of a room. There we made 5 measurements and reported the mean as 5.828 meters. If we had wanted to express the same measurement in millimeters instead of in meters we could have reported 5828 mm. Remember, now that 5828 mm is the average of 5 measures of which the lowest (shortest length across the room) was 5823 mm and the highest was 5834 mm. This spread from lowest to highest we call the "range" of the 5 measures. We subtract and express the range as 11 mm.

Our mean of 5828 mm is 6 mm below the top of the range and 5 mm above the bottom. We may want people to know the range, too, when we report the mean of several measurements so we use 5828 ± 6 mm. Or we could use the meter unit and report 5.828 ± 0.006 m.

This system, using the mean and expressing the range, is an overly simplified statement of the method, but it does show how the system works.

There is yet another way to express the uncertainty of measurements. As a simple example, let's say that you are measuring a rod with a meter stick and find that it is just under 1 meter; it is 0.999 m. That is, you wish to report that it is nearer 0.999 m than to 1.000 m or nearer to 999 mm than to 1000 mm. You are saying that the range is within 1 mm out of about 1000 mm. Therefore you report your measurement as 0.999 m with an error of 1 part per 1000.

Suppose, however, that you take the same meter stick, measure a room that is close to 10 m across (or close to

10,000 mm). You find that several measurements range from 9995 to 10,005 or a net range of 10 mm. This, too, shows an error of about 10 parts in 10,000 or 1 part in 1000, as in the previous example.

To say that these measurements are accurate to 1 part in 1000, we must assume that your meter stick is "exactly" 1 meter—which of course we cannot do because we are working on the theory that no two things are exactly alike (including your meter stick and some master meter stick). All measuring instruments, including wooden meter sticks, do change from their original dimensions. This part of our measurement problem—the unstable nature of things—the changes that come with temperature, moisture and other factors—must always be expected as part of our measuring and must be allowed for if we are trying for high accuracy. Scientists working on one of the most important measuring problems—the measurement of the speed of light—and using the best equipment and methods they can devise, must still allow for the uncertainty of their measurements. Thus, they report the velocity of light in vacuum as 299,792,456 $\pm$ 1.1 meters per second.

We have discussed three different ways of reporting the uncertainty of measurements: (1) reference to ½ of the smallest scale division, (2) reference to a range of several values above and below the mean of those values, and (3) the error in parts per 1000 or parts per million, or the like. No matter which system you use for reporting the uncertainty, it, the uncertainty, is there even though scientists have worked hard at finding ways to evaluate the uncertainty in any particular measurement.

Evaluating Results

When you listen to one song by a performer and like it and go and buy a recording to hear more songs by that performer, you have taken a sample and have generalized from the sample to the larger group from which the sample came. We do this sampling and generalizing many times a day in informal ways. Scientists have become more formal about it. They have worked out systems of logic and statistics to help in judging how well a sample represents a larger group, usually called the *population*.

The example of five measurements across the room in the measurement chapter may be thought of as a small sample of all the possible measurements that could be made. We could set up a program of making many such measurements—of many people each making many measurements—so that one might eventually have thousands or millions of measurements.

The large, unknown number of measurements of which any one measurement is considered a sample can be known only from the sample. This is like eating a cookie from a cookie jar. No matter how enjoyable the first, second or third, we never know how good the remaining cookies are from the samples only. We can only predict or infer that the rest—the population from which the sample came—are like the sample. But we can never have all the measurements of a dimension of a thing. That is, we could in theory keep on making measurements forever (assuming the thing didn't wear out, and we didn't too!).

How does a scientist judge that his sample of measurements (or findings expressed other ways) fairly represents all possible measurements? Let's say that Ken is doing an experiment as his science project in which he has planted corn seeds in two planters, to test the value of a fertilizer. He is going to compare the two plantings, one with the fertilizer and one without. At one stage he measures the plants in one planter, all growing under uniform conditions. To keep the numbers small for quick, easy working with them, let's say that the planter has five healthy, growing corn plants. Suppose that Ken measures the heights of his five plants and finds the following:

Plant A	57.2 cm
Plant B	57.2 cm
Plant C	57.2 cm
Plant D	57.2 cm
Plant E	57.2 cm

What? All the same? Most of us who have had experience with growing things would say, at once, that this is highly improbable; that it is just a thing of chance the plants would all be so precisely the same height. Correct! It is a matter of chance or probability. Probability, you will find, is the main theme in a study of the evaluation of scientific findings. Suppose, now, that Ken's measuring had brought the following results:

Plant A	57.9 cm
Plant B	55.7 cm
Plant C	58.4 cm
Plant D	59.2 cm
Plant E	57.3 cm

"That's more like it," we would say. We expect differences in things, especially in living, growing things. That is, it is more highly probable that there would be such differences and not so much sameness.

Now, whether we like the sample or not, it is *all we know* about the larger population of plants that Ken's supply of seed might grow. Suppose, again, that someone planted a hundred seeds from the same supply as Ken's and under very much the same conditions. Then suppose he went to work measuring them at the same stage as Ken's plants. We would like to see how the sizes vary in this much larger sample, so we make a *frequency distribution* (See Figure 10.1) showing the sizes. That is, an "X" mark is made for each corn plant over its height measurement, listed along the bottom of the chart:

Figure 10.1
Frequency Distribution of 100 Corn Plants

Number of corn plants at each height	50	51	52	53	54	55	56	57	58	59	60	61	62	63	64	65
14																
13																
12								x								
11						x		x	x							
10						x	x	x	x							
9					x	x	x	x	x	x						
8					x	x	x	x	x	x	x					
7				x	x	x	x	x	x	x	x					
6			x	x	x	x	x	x	x	x	x	x				
5			x	x	x	x	x	x	x	x	x	x				
4			x	x	x	x	x	x	x	x	x	x	x			
3		x	x	x	x	x	x	x	x	x	x	x	x			
2		x	x	x	x	x	x	x	x	x	x	x	x	x		
1	x	x	x	x	x	x	x	x	x	x	x	x	x	x		x
	1 plant 50 cm high	3 plants 51 cm high														

Height, centimeters. (Rounded to nearest centimeter)

We see that there are not many of the shortest plants and not many of the tallest but more of each size in the middle of the range. If we drew a line over the tops of the columns of sizes and if we had many more specimens measured and recorded, the lines, or line graphs, would look something like one of these:

Figure 10.2

Normal Distribution Curves Come in Many Different Shapes

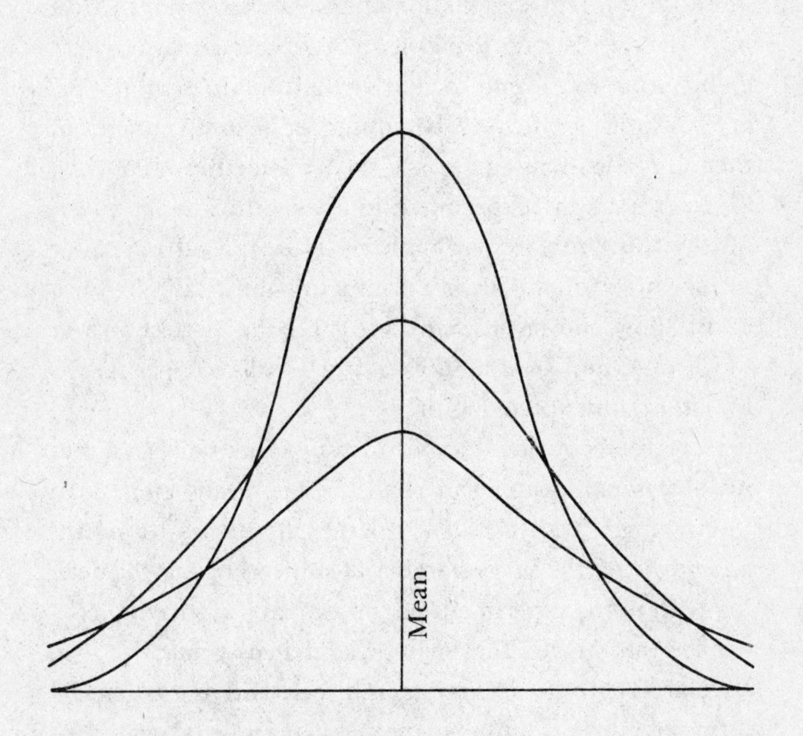

The curve (or line graph) represents the variability of the distribution around the mean.

Such a distribution of a large number of things (and it must be large—preferably thousands) is called a *normal distribution.*

Many things when they are measured and graphed like this show normal distributions—things such as the heights of people picked at random (in large numbers), the amounts of food eaten per person per year, etc. This widespread nature of things to show normal distribution has been used by scientists and statisticians to work out ever more meaningful designs for science investigations. Most modern scientists are thinking about the statistics they are going to use in the end to analyze their findings while they are working at the very beginning or planning stages of their investigations. They are saying something like this: "I don't want my experiment to come out as some queer, quirky thing that proves nothing. How must I plan now so that my results will be meaningful statistically?" A scientist knows, however, that there can be no perfect answer to his question. He can always, just by chance, get results that show unexpected quirks.

Nevertheless, as he does his investigation, he is trying to uncover some meaningful results. This means more than just saying, "Yes" or "No" to the hypothesis. It means going beyond the small number of subjects he may be dealing with in his experiment or survey. It means having confidence that his findings may be stretched–generalized–to any larger group of similar subjects. Did ingredient *Q* seem to prevent sunburn in the experimental group of people who used it? If so and if that experimental group fairly represents the larger population, we may then reasonably expect that ingredient *Q* will prevent sunburn in most of the larger population.

The use of random choices in the first stages of an investigation means more than just helping to keep the scientist's prejudices from affecting the results. It helps to assure that the sample (of people or other subjects) used in the investigation will allow us to generalize to the larger group that the sample is intended to represent.

Can You Prove It?

In his corn growing experiment, in Chapter 10, Ken used a controlled experiment design; a control group of five corn plants and an experimental group of five. To the experimental group he added a chemical fertilizer–urea, a nitrogen compound that may be put into the soil or dissolved in the water given the plants. His independent variable is the addition of the urea to the experimental group. His dependent variable (if he observes one) is the difference in growth rate (height) of the plants in his two planters.

At a proper time in his experiment he measures the heights of the plants with the following results:

Figure 11.1

Heights of Corn Plants in Ken's Experiment

	Control Group (*No Urea*)	**Experimental Group** (*Urea: 2 grams per liter of water*)
	57.9 cm	60.3 cm
	55.7	60.0
	58.4	57.6
	59.2	61.2
	57.3	58.9
Totals	288.5	298.0
Means	57.7 cm	59.6 cm

Difference between means: 59.6 — 57.7 = 1.9 cm

We see that there is a difference between the means (commonly called average) of the two groups. The difference is 1.9 cm in favor of the experimental group; the plants of that group average 1.9 cm taller than the plants of the control group. This looks good.

"See!" Ken says. "Adding urea to the experimental planting has made the corn grow faster." Can he be sure of this? No, he cannot. Maybe it was a chance happening that he got five taller-growing plants in the experimental group and five shorter-growing plants in the control. He should not make any decision just yet. He should get someone to make a good statistical treatment (unless he can do it himself!) that would go beyond comparing the mean heights of the two groups.

A statistical analysis would show how much the heights vary among themselves. Then it would show how the means compare with a larger "population" of plants like Ken's. Where would this larger population be found? It would be imagined, inferred, hypothetical; it would be created out of the variability, the range, the scatter of his sample—and the size of the sample. It would be created by the use of equations to be found in statistics books.

Further, then, a judgment would be made about the chance—the probability—that the difference Ken found was, or was not, simply a chance difference. This, too, would be done by reference to appropriate tables in statistics books. Actually, Ken's findings would be supported by agricultural research by professional scientists and by the experiences of thousands of farmers who have found it useful to apply urea and other nitrogen compounds to their corn plantings.

With all of this support, why wouldn't scientists flatly say that they have proven the value of this treatment of corn? The problem lies partly in the question: How can you know when you have proven a thing to be true? And it lies partly in the way the words "prove" and "true" are used in mathematics and logic—and in ordinary speech.

First the mathematics and logic. You and I can agree that this is a true statement in arithmetic: $148 + 293 + 167 = 608$. We can prove it by doing the addition. That is, we follow certain rules of mathematics to prove the statement is an equality. Mathematicians would not agree, however, that we had proven it by simply doing the addition. They would go at it in a roundabout, more complicated manner. But, in the end, they would show that the statement was proven by agreeing on certain things about arithmetic and its rules.

In logic of the formal sort, proof would be much the same, as in this example:

If all wangtups have gitly speekrongs,

And if Q is a wangtup,

Then Q has gitly speekrongs.

Even though the statements do not mean anything in real life, yet if we accept the first and second statements as true then the conclusion, the third statement, is true. The "proof" is all there in the statement. It has nothing to do with real people or things and their mixed-up ways.

These simple examples do not, of course, do justice to mathematics and logic. Both are fascinating and powerful tools of thought or reasoning that man has created. Their proofs, however—their truths—are so much different from the kinds of proofs that scientists are seeking that it becomes awkward to try to use the same language. And, since mathematics and logic got there first with the terms "prove" and "true," scientists in recent times have pulled away from using those terms.

In ordinary experience, too, there is a problem with those key words. Most people would say, "See, Ken proved it! It's true that urea makes corn grow faster." Or they might say, "That proves it! Hocus is better for a headache than Pocus," even though they may have used the remedy only once and the test has serious weaknesses. Or, again, "That proves that dreams do tell the future! I knew you were coming because I dreamed about it!"

These difficulties with the language do not provide the main objection, however, to the use of "prove" in scientific work. When we talk about "proving" something in science we are, in effect, predicting the future as well as examining the present. How much can we depend on something happening in the future just because today's scientific findings show it probable now?

In Ken's experiment, for example, he used only five plants in each planter. This is a small sample. It cannot tell us much about the larger population of future corn plantings, no matter how much statistical analysis we apply to it. However, let's do some more analysis of Ken's results to see how this helps to learn about the predictive value of his findings. Let's rearrange the measurements of the corn plants according to height. We will rank them as in Figure 11.2:

Figure 11.2

Ken's Experiment: Two Groups Compared by Ranking

	Control Group *(ranked shortest to tallest)*	**Experimental Group** *(ranked shortest to tallest)*	**Difference**
	55.7 cm	57.6 cm	+1.9 cm
	57.3	58.9	+1.6
	57.9	60.0	+2.1
	58.4	60.3	+1.9
	59.2	61.2	+2.0
Totals	288.5	298.0	+9.5
Means	57.7	59.8	Mean Difference +1.9

Does this tell us more than simply comparing the means? Suppose his results in the experimental group had been as in Figure 11.3 (also ranked by height).

Figure 11.3

Ken's Experiment: Greater Variability

	Control Group	Experimental Group	Difference	
	55.7 cm	53.6 cm	−2.1 cm	−2.5
	57.3	56.9	−0.4	
	57.9	60.0	+2.1	+12.0
	58.4	62.3	+3.9	
	59.2	65.2	+6.0	
Totals	288.5	298.0		+9.5
Means	57.7	59.8	Mean Difference	+1.9

Here we see that the difference between the means of the two groups is the same as in Figure 11.2. But notice the range of heights in Figure 11.3. The experimental plants are not as uniformly taller than control as they were in Figure 11.2. There is more variability. These results would provide a less reliable basis for predicting about future plantings.

I hope that you begin to agree (if you had not already

known about this) that statistical treatment of data can reveal useful information. Finding the means and their difference is statistical analysis. Ranking the heights and comparing the pairs of plants is statistical analysis. These two ways of analyzing data are very elementary (even antiquated) as compared with the methods used by people with more mathematical and statistical knowledge.

How could Ken "prove" more, besides just making statistical analyses of his data? He could do it again—*replicate* his experiment. This would raise the predictive power of his test if results were as good as the first test or better, even though it would still not finally prove anything. We must accept this because there is always uncertainty about the future. Some things are more highly probable than others, of course. We are all fairly sure that the sun will come up tomorrow, while we may not be so sure that another planting of corn, treated as Ken's was treated, will turn out the same. So, you see, we are always dealing in probabilities.

Scientists like to show that their findings allow them to predict, or generalize, in another way than in the simple replication of an experiment or other investigation. Ken could expand his research in several ways:

Plan A: One experimental level of urea, applied in water (Ken's first plan).

Planter	*Description*
1	Control; no urea.
2	Experimental; 2 g (grams) per liter of water used to water the plantings.

Plan B: Three experimental levels of urea, applied in water.

Planter	*Description*
1	Control; no urea.
2	Experimental; 2 g urea per liter of water.
3	Experimental; 4 g urea per liter of water.
4	Experimental; 6 g urea per liter of water.

Plan C: Three experimental levels of urea, applied in soil.

Planter	*Description*
1	Control; no urea.
2	Experimental; 10 g urea mixed in the soil.
3	Experimental; 20 g urea mixed in the soil.
4	Experimental; 30 g urea mixed in the soil.

If he were to test both variables, two ways of applying urea and three different levels of urea, at the same time he would need an arrangement of planters (or outdoor plots) as in Figure 11.4.

Figure 11.4

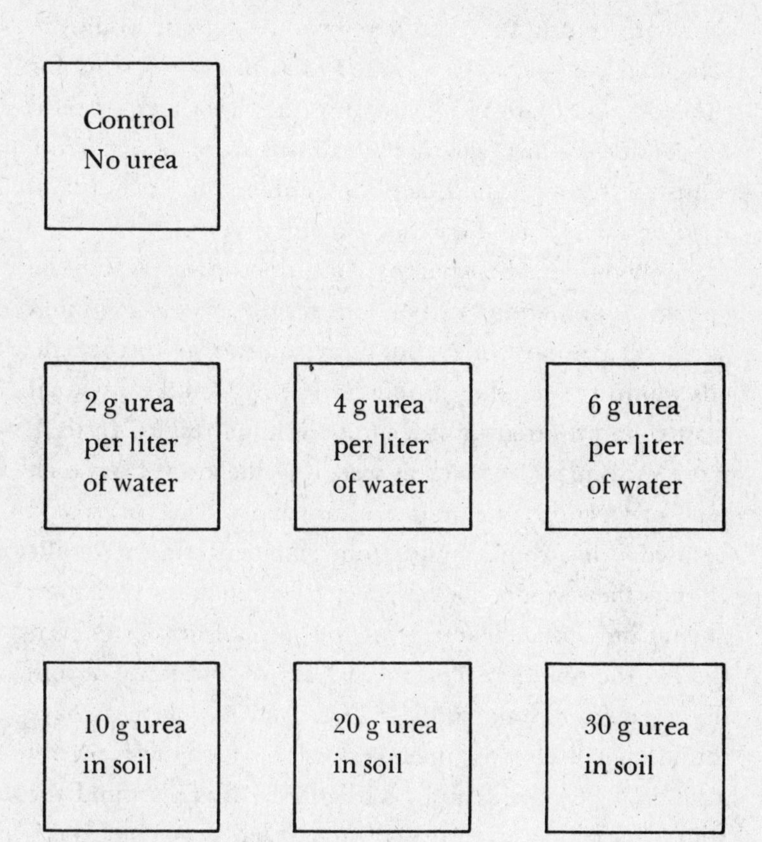

You may be interested in figuring out how many different experiments would be needed to test each of these one at a time against a control and one at a time against each different level of urea. Then recall that there are other nitrogen compounds that should be compared with urea and should be applied in different amounts, other kinds of soil, other varieties of corn, other planting methods and methods of applying the fertilizer, other chemicals that may be as important as nitrogen for healthy growth of the corn. Many of these variables would best be tested in combination with certain others. Therefore, the designs would, in some cases, be more complicated than in Plan C. And for the most significant results, most of the experiments would be conducted through to the mature stage of the crop. Thus the testing would be done outdoors in plots of land large enough to handle with farm equipment.

Surely under these larger conditions there would be enough "population" to make the results prove something! Well, perhaps not *surely,* but more so. And yet these methods would create other problems. Rarely would individual plants be measured in order to determine results. Instead, more gross measures such as weighing the grain from each plot, or weighing the grain and other plant matter, would be used. This would increase our confidence in the results in that they would not be affected so much by variations among individual plants as in Ken's small groups. Nevertheless, the different plots might vary as to quality of soil, drainage conditions and the like. And so scientists have found that each "treatment" must be used over several smaller plots spread in a randomized pattern around the whole field. That is, instead of two larger plots, one experimental and one control, the larger experimental plot is divided into, say, five smaller experimental plots (each with the same treatment) and the larger control plot is divided into five smaller control plots. These are distributed randomly about the whole area. And so we find that we are dealing with a small number of things (five plots) just as with Ken's experiment (five corn plants). While this gives important improvements in the overall plan, it still shows somewhat the same problem of a small sample (small number of plots). And the statistical treatment must be advanced, highly developed, to squeeze the most meaning out of the results.

With it all, there is still this uncertainty as elsewhere in science. We must not be disheartened about the uncertainty, however. Unfortunately, many people have been oversold on science and its powers for finding out the "truth" about things. Others have shown disappointment that science has not been able to solve more problems. It is important to see that scientific methods are the best that have been found for learning the truth about many things —and are superior to ordinary, everyday "common sense" methods. That's why scientific methods are called "scientific"; they are better than unscientific methods. Yet mankind has been working with scientific methods only a short time, as compared with other methods. Not all kinds of human problems can be solved with scientific kinds of knowledge. And those that might be solved by scientific methods seem to be unlimited in number. We must honestly admit that scientists probably will never get to all the problems that they might investigate and help to solve.

Nevertheless, in spite of the uncertainty of science and limited speed with which scientists can move into new areas, we must use scientific methods to find out all we can about the world and the people and things in it. Even with its uncertainty it is the best we have for many of the problems of mankind.

How to Get Help in Evaluating Results

How serious should you be about evaluating the results of an investigation?

Is it possible not to evaluate? Not to judge?

As soon as you see the findings of a study, you probably start making a judgment about it. So as long as you are going to evaluate you ought to make as good an evaluation of the results as you can. Perhaps you do not know how to do statistical treatment beyond the simple finding of the means of two groups to compare, or the ranking of measurements of specimens (as in Ken's experiment). Or, perhaps all you can do on your own is to put the data in tables so that relations between variables can be easily seen and informally compared. Do those things of course; but if more seems needed, get help. There are people who may help, and books. (Books are other people, of course, but more remote.) First try the people near at hand.

If someone helps you do a "professional" statistical analysis of as simple a set of data as Ken's experiment, it may take an hour of time or more, depending upon how fluent the person is with statistics, and whether or not your work must all be done with pencil and paper or there is a calculator available. This time-consuming part of it you must consider when you ask someone to serve as your counselor in statistics—your statistician.

If you have not gotten as far as high school algebra, you will have to depend altogether upon others to set up a plan for processing your data. However, the grubby part you should do—the adding and other arithmetic. Ask for help first from teachers in your school. You may find one who has had a college course in statistics and who would like to brush up on its use. Most teachers, however, usually have little practical need for statistics even when they have studied it and just may not want to spend the time necessary to freshen up their knowledge in order to help you. And many, of course, have not had college work in statistics; it simply may not have been part of their college program.

Consider asking another student who is farther along than you in mathematics. You may find a high school mathematics "major" who would like to do a little digging in statistics in order to try his mathematics skills on a real problem. Perhaps a mathematics instructor would help you or would put you in touch with a math student who would help. And then, again, there may be professionals in the community outside of school who are using statistics in their work and who would help you—professional scientists, statisticians, engineers or other technicians.

On your own, you should move into statistics as far as you can go. You should get whatever books you can find in the school or community libraries on design of experiments, surveys, statistics, etc. Even if a book, on the whole, is beyond your grasp, some reading of the introductory material may prove profitable. It may help you begin to grasp the language of statistics and to see how varied the patterns are. You may be surprised to learn that new statistical methods are constantly being developed. As different scientific methods are created for different projects, it becomes necessary to invent statistics to go with them.

In statistics, as in most other human affairs, there is disagreement among the professionals as to which statistics are best for a particular research design. Some prefer the kinds of statistics based upon the "normal distribution" and consider them perfectly proper for many, or most, analyses. Others disagree. They argue that there is a weakness, especially for small samples, in the normal distribution kinds of statistics. They say the weakness is something like this:

1. You use the sample to imagine (or to infer or to postulate) a "population" from which the sample was supposed to have been taken.
2. Then you compare the population with the sample (in the statistical analysis) to judge whether the sample was just a far-out quirk or was more regular.

The critics say that this is circular reasoning, or reasoning in a circle, which is not good logic—that is, using *A* (the sample) to set up *B* (the population), then using *B* to judge *A*. Such critics of normal distribution statistics prefer to use "non-normal" or "non-parametric" methods.

As a beginner, you cannot be too concerned about arguments among professionals. But a hint at their controversy will help you to see that mathematics (including statistics) is not a standstill sort of thing. New mathematics is constantly being created to fit the needs of people as the needs change.

In your effort to get help with your statistical treatment, you should try to get your consultant in on the planning stages of your project. Design of the investigation and statistical treatment should go together. It is altogether too easy to make an elaborate investigation and reveal nothing significant. Remember the project for testing running speed as related to loss of sleep? Suppose that the experimenter had never heard of the counterbalancing plan and, of course, did not use it. No matter how striking a difference between the running records of the two groups, the results would show nothing believable without the counterbalanced trials.

Of course, if you are to do as simple a project as making a measurement of the boiling temperature of water, you will not need much statistical treatment of the data. However, if you make several different measurements at different times and try to do them independently (so that each new test is not influenced by your previous measurements), you may find considerable variation. You may then have something of a problem evaluating your results. You will want to consider questions like these: How much of the variation may be traced to differences in the water? In the apparatus? In other surrounding factors, such as atmospheric pressure? Your effort to find relations such as these will probably lead into statistical treatment of the data.

Or if you intend to find a relation, say, between the boiling (or freezing) temperature of water with different amounts of a salt dissolved in the water, you will almost certainly want to graph the results and perhaps get some statistical counseling.

If you do a survey of the simple counting or census type, you will need less challenging statistical treatment than if you try to find a relation between two variables, such as eating a good breakfast and performance on school tests.

How about evaluating the results of naturalistic observation? In the observing of people and things as they naturally proceed, you may accumulate large amounts of data. This probably will not permit much statistical treatment, but would adapt, rather, to tables and graphs along with word descriptions. Your evaluation will have to come from comparing the results of your naturalistic observation with the results reported by other skillful observers.

As you plan your project, state as clearly as you can the question or problem you will investigate. Do this in writing. Then write out a description of the method you plan to use. As you plan method and materials, you will see more clearly how to express your hypothesis. It, the hypothesis, should state as clearly and precisely as you can what it is that you expect to show or test.

Example A.

Question: Does nitrogen help to make corn grow better?

Hypothesis: Sweet corn planted in sandy loam in planters in the laboratory and treated with urea in the water supplied to the plants (2 grams urea per liter of water) will show a significant increase in height at two weeks after emerging as compared with a control planting.

Example B.

Question: Can mice grow normally on dried peas as the only food?

Hypothesis: White mice born in a cage and fed only whole dried peas will make normal growth to age 60 days as compared with a control group fed Brand *K* mouse food.

Your hypothesis, it is hoped, can be stated definitely enough that it can honestly be tested. But proven? No.

The actual doing of scientific research is the work of most scientists. But there are some scientists and others whose main concern is trying to show what can be proven and what cannot be proven by scientific methods. Actually in science there are two very different activities: (1) the active research, and (2) the thinking about what it means to "do" science and to do it "right." The people who do the second activity are, primarily, philosophers; that is, philosophers of science. They may be scientists also, of course, but they are working in an importantly different capacity when they are concerned with the reasoning, the logic, behind scientific discovery and explanation.

As the philosophers of science have explored more deeply into what can and what cannot be proven in the work of scientists they have come to agree on this view, substantially: If you have a hypothesis that is expressed as a positive statement and which is testable by scientific means, you are limited in what you can prove about it; limited in the sense that you cannot prove it true but that you can prove it false. Take as example the hypothesis about mice growing normally with dried peas as their only food. You test the hypothesis on a group. The mice all grow normally. Have you proven the hypothesis? Not necessarily; no matter how many tested groups grow normally, the hypothesis is not permanently proven. If only one mouse fails to grow normally in any future test, the hypothesis is disproven. That is, the hypothesis may be disproven but not proven—at least not permanently.

Professional scientists have learned to live reasonably comfortably with this kind of logic or reasoning. They have learned to live with the uncertainty that seems to be built into human knowledge—to live with the uncertainty but to do their best to reduce the uncertainty to something manageable.

So, plan to do your project. Do what you can to get help in judging the outcome of your research; try to have the help—your "statistician"—in on the planning stage. If you can't, do your project anyway. Get the results even if you may not be able to analyze them as fully as a professional would do. Maybe, someday, you will be able to come back to your report, analyze the results further and say, "Well, see what I discovered—I think!"

Reporting to Others

Belief is our problem; what to believe? And why? In many human activities people are expected to believe, with no doubt about it! One who does not believe is out! In science it is just the other way. Everyone is expected to doubt. Every existing theory, principle, law, is expected to be changed in time as new findings come to bear. Oh, yes, among scientists there are believers—those who are so convinced of the rightness of a certain principle that they just do not admit doubt. But modern scientists, on the whole, have accepted change as ever with us. They have accepted, in principle, also, that we never know the truth about the universe or anything in it. The best we can do is to keep working on our methods of discovery. As we change them or improve them, the "truth" will change and perhaps become more truthful.

Since modern science depends so much on getting others to agree on the evidence, does this mean that scientists are an agreeable bunch of people, on the whole? No; controversy is plentiful. There is controversy over methods—which methods yield the best results. There is controversy over theory. A theory is a hypothesis built large. Can anyone *know* that dinosaurs looked like that? Or existed then? Or became extinct for those reasons? No; of course not. It is all theory and very likely will remain theory, since we have nothing like time travel to permit us to get direct evidence. And, so, the dinosaur theories and many others will remain theories until (if ever) ways are found to test them by empirical means, by testing and observation.

Nevertheless, scientists many times do not stick close to the evidence. They like to reason, often, from present evidence into the past or into the future or into unexplored "presents" and tell what they think. Also, society often demands this of scientists—demands that they predict beyond current findings. This predicting (or extrapolating beyond the actual findings) guarantees argument.

In spite of all of this controversy and disagreement, modern science is a science of consensus, of agreement. You publish your findings. If other competent workers cannot disprove your findings, you are in, no matter who you are. There are no requirements of age, sex, wealth, race, etc. It is evidence that counts.

To present your evidence, you must be able to make a report that is good in detail, that permits those who read it to find in it enough information that they may judge the validity of your findings and, if they wish, to replicate your investigation. The recommendations following are general. You must make your report fit your investigation as best you can.

1. Introduction

State the question or problem you have tried to answer or the theory you wish to test. The word *theory* is used to mean an untested principle, and so we have relativity theory, the "big bang" theory, the theory of evolution, etc. Once a theory has been sufficiently tested, validated, "proven," it is no longer a theory. It is then a principle or law (until, of course, it is replaced by a better principle or law!). Your title may also state the question or problem but more briefly than here in your introduction.

You may wish to put an "abstract" at the beginning of your report. This is for the benefit of the reader who is not sure that he wants to read your whole report. It tells the story in perhaps 100 or 200 words.

In your introduction you may want to tell what brought you to make this investigation; that is, what other explorations brought you to this (not, of course, that you want to get a good grade or to win a prize!).

State your hypothesis. This is narrowing down your goal. You take off from a general theory or question or problem and try, in your hypothesis, to be as specific as you can in stating what it is you are (or were) trying to test. Our current system of science is generally called a hypothetico-deductive system. That is, scientists would like to test, usually, a broad, general theory. However, they cannot (or rarely can they) test the whole theory at once. They must, by a process of reasoning, or logic, deduce from the theory a smaller, local, testable hypothesis—a statement with which they can come to grips by available techniques and materials. It may take many tests of many hypotheses to validate the theory.

So you state your hypothesis, or more than one if your study included the investigation of more than one. You should plan to restate the hypotheses in your report, as necessary.

2. Procedure

Here you describe methods and materials. You must think always of your reader as not being present to see the things you used and how you proceeded. What will he need to know about this? Could he take this description and go with it—replicate the study?

In the discussion to this point, I have said little about record-keeping. Your records, of course, are the foundation of your report. If you were to work in secret with no thought of ever reporting to others, your record-keeping would be altogether your own concern. However, if you want others to recognize what you have done, your records must always be designed to help you in reporting to others. The records, themselves, should bear inspection by anyone. They should tell anyone reasonably familiar with the kind of work you did all he would need to know about your investigation.

Do not depend upon remembering things. Keep a diary or diarylike notes. Each time you do something write it down or dictate it to a tape recorder for later transcription, record it in a chart or table, make a sketch of it or a diagram or take a photograph of it. And always, always, every time, date your record, including clock time if this may be important. For most of your diary-record use a bound notebook (not loose-leaf), which lies flat when open. Write with pen (not pencil). Do not expect to erase. If you need to change a statement, draw a single line through it but leave it so that it may be read and rewrite the statement. It is best to write double-space in the notebook.

It is perfectly good procedure to write ideas for other investigations in the diary-record as you go or thoughts about how you might have done this better—or thoughts about how well you did do this! You see, you must think of your record as a rich source of material for your report.

It will contain a great amount of material that you will not put into your final report, but the material should be there if needed. Be generous in your record-keeping.

If, in your absence, it should be necessary for another person to make observations for you, be sure that he is well-trained in the work and make certain that he will do the record-keeping at least as well as you would. Identify him and his part of the work in the record.

If you find it necessary to make records separate from your notes—records such as tape recordings, diagrams, photographs, charts, etc.—number them and date them; refer to them in your notes so that they are tied in to the chronological record (your diary).

All of these original records must be kept. Do not copy the originals and then throw out the originals. Keep the original records. It is so easy to make mistakes in copying. Then, too, there is a temptation to edit in the copying—to change something so that it looks better. If you need a copy, make it, of course, but KEEP THE ORIGINALS!

All of these record-keeping requirements! Why? Remember that you will forget. Important details may become vague after a while. Or someone may ask a question about your work that you may not have thought about. Your records may help you answer. And (who knows?) a hundred years from now someone may be saying, "See! Here is the original record of one of his (or her) early investigations, done long before the fame of later years. Isn't it nice that we have these?"

3. Results

In your notes, you will have recorded both method and results in a blend of records of various kinds. However, in the writing of your final report, you will probably separate the two into a methods section and a results section.

The methods will be mainly word descriptions plus diagrams, pictures, etc.

For results, if you recorded only a small amount of data, facts, measurements, you may just put all of the raw data in your final report. If, on the other hand, you have mountains of raw data, it is unlikely that the whole of this should appear in your final report. Rather, it should be summarized in graphs and tables of various kinds or by whatever means is best suited to your goal of presenting clear and convincing evidence.

The calculations of your statistical analysis need not be shown in full. You must show the results of the calculations and identify the methods. If you report a mean, or average, of 49.8 "everybody" knows what you mean. However, if you report a correlation of two series of measurements, you must tell which correlation method you used; in fact you must explain your method with most statistical treatments.

How you judge the validity of your results will depend largely upon the extent to which you get into statistical analysis. Any statements you make will grow out of your understanding of the statistics or will be made with help of your consultant in statistics.

In all of the above suggestions for making a report, there may be conflicts with other requirements. Your instructor in science may have different requirements for your report if you do a project under his guidance. You should comply with his standards, unless, of course, you can convince him of the desirability of doing things differently. Then, too, if you get so far as publishing in a magazine or journal, the editor will have ideas about how to present your report. Even though you may not get to publish in a journal, you may find it interesting to go to a college or university library, to find journals in your field of research and see how the professionals are currently making their reports.

Much of this kind of reporting is dry, factual and, in fact, not interesting, unless you are deep in the subject. Yet, occasionally, a science paper turns up with style, wit, grace. And at no conflict with the content! What a pleasure! If you have found that you can write in an interesting style, do try to apply your style to the writing of your science report.

Glossary

Anecdotal evidence. Evidence given out of one's experience, told as a story or anecdote. As evidence, usually judged to be lower quality than scientific evidence.

Average. See "Mean."

Bias. Leaning away from some expected normal or norm or standard. Sometimes used as a synonym for prejudice.

Blind experiment. In a controlled experiment with people, it may be important that the subjects (the people being tested) *not* know whether they are in a control group or in an experimental group. They may, and should, know what the experiment is about; that is, they should know what the independent variable is. They should know what the experimenter may do to them or with them and should have agreed to that treatment. However, in a blind experiment, the individual subject-person is not told whether he is actually getting the treatment or is in the control group NOT getting the treatment. See "placebo."

In a controlled experiment with animals we assume that the individual animal does not know that he is getting a certain treatment that should have a certain result. Suppose it does know! One should watch this, as experimenter.

Some people believe that plants can sense the mood of the experimenter or otherwise react to planned treatment. So?

Case study. The observation or study of an individual or an event more or less by itself—although the term may be applied to one of a larger group or series.

Classification. The separation of things into groups because of likenesses and differences. Things are put into Class A because of likenesses that are different from the likenesses that cause things to be put into Class B.

Common sense. Beliefs held by many people, by the "common man," common beliefs; but, however, beliefs that have not been tested in ways that would satisfy those who insist upon scientific methods for testing.

Confounding. In an experiment, the experimenter would prefer, usually, to have only one independent variable. If other variables enter the experiment without his knowing of them, their effects may hide those of the planned independent variable. Or he may credit to the independent variable effects that may more properly be credited to the (unknown) other variables. Confounding is not all bad, however. If the experimenter plans more than one independent variable into the experiment, he may be able to use statistical treatment of the results that will show the (probable) separate effects of each.

Consensus. Usually refers to agreement of an informal sort, without "putting it in writing" or otherwise stating it exactly.

Constants. In an experiment, constants are conditions that might vary, in other times and places, but that the investigator tries to keep from varying for the purposes of his investigation. Constants must be contrasted with an independent variable, which is made to vary (by the experimenter), or a dependent variable, which is allowed to vary. Constants may be established as part of survey methods, also.

Controlled experiment, control. In a simple experiment, there is one subject (or group), only, that gets the experimental treatment. To judge the results better, the experimenter may use another subject (or group) that does not get the experimental treatment but is, otherwise, treated like the first. Thus, there is an experimental subject and a control subject (or "control") in a controlled experiment.

Counterbalanced experiment. (Also see controlled experiment.) A controlled experiment may be done twice with the same subjects; on the second time through, the control becoming the experimental and the experimental becoming the control for the counterbalanced design or cross-over design.

Data. Information, facts, figures growing out of an investigation and put on record. Not limited, of course, to written or printed records. Data may not be correct (or "true"), yet may exist as data. "Data" is both the singular and plural form although sometimes "datum" is used as singular.

Dependent variable. In an experiment the experimenter changes something to observe what happens. The things he changes may "cause" something else to happen. If it does, the "something else" is called a dependent variable. That is, the "something else" happens only after the occurrence of the main change planned into the experiment by the experimenter. So, the something else *depends* on the main change (or independent variable) and is therefore called the *dependent* variable.

Double-blind experiment. (See "Blind Experiment.") If the experimenter in a blind experiment cannot trust the subjects to be honest about their reactions to the experimental treatment he arranges so that no subject knows whether he is or is not getting the planned treatment (the independent variable or experimental variable). Now, it may be that the experimenter must judge the effects of the treatment. Suppose that he feels that he cannot judge honestly; he is prejudiced. He then plans to hide from himself, as well as from the subjects, which of the subjects is getting the treatment and which is not. He goes on with the experiment and judges the results for each subject. Then he allows himself to find out which subjects got the treatment and which did not. Thus, in keeping himself and the subject uninformed until the end about which got the treatment and which did not, he has made it a double-blind experiment.

Empirical. Refers to the testing of a statement, question or problem by scientific methods or other practical, real-life methods instead of testing by reasoning or logic only.

Evaluation—evaluating results. The attempt to decide whether or not a scientific test has shown a satisfying or usable result, usually expressed in terms of a probability. An attempt at answering the question: Did the investigation work out to be a good test of the hypothesis?

Experiment; experimental. An investigation made by changing things, manipulating things (including people, on occasion), to find out what happens as a result of the changing of things. An experiment is usually planned to answer a specific question, solve a problem, or to test a statement or hypothesis. Distinguished from unplanned change or just "letting things happen naturally."

Experimenter. One who plans and performs an experiment or, at least, one who applies the independent variable and who observes the results.

Extrapolate. See "Predict."

Frequency Distribution. A method for classifying data (measurements, countings) in which, usually, ten or more groupings are made and the number, or frequency, of items in each group is shown.

Glossary. I like to think that no word can be completely defined by one person for another person, and that no two people can have the same meanings for a word, no matter how hard they try to use words to define a word (including dictionary writers and writers of glossaries). In this glossary, then, I suggest meanings that may help in the reading of this book. These are limited meanings —more limited than a professional work or a good, general dictionary would give. These are glosses (for which meaning, see that good, general dictionary).

Hypothesis. In a science investigation, the hypothesis is the statement that is to be tested, a statement of the purpose of the investigation, or what it is the investigator intends to "prove," or to disprove. (Plural, hypotheses.)

Independent variable. First, of course, a variable is a thing that shows change. A person doing an experiment changes something to observe what happens. The "something" that he changes is the independent variable.

Logic. Efforts to prove something by making statements or "arguments." Not all logic is good logic, of course, in that many times an effort to prove something by statements does not succeed. In the planning or designing of an experiment (or other investigation) one uses logic, arguing that the design of the experiment will, when carried out, show the hypothesis to be valid or not valid.

Mean. An example: Of the three values, 8, 10, and 12 one finds the mean (or average) by adding the three, total 30, then dividing by the number of values (3), to get the mean: 10. Algebraic formula: $M = \frac{\Sigma x}{N}$ where M is the mean; Σ (capital sigma) means the operation of adding all the values or the "sum of X's"; X is a value or score (of which there are 2 or more); and N is the number of values or scores. (Also see Figure G.1).

Naturalistic observation. To observe things, animals, people, groups, societies as they exist and regularly "do their thing." The plan is to disturb the thing observed as little as possible, recognizing all the while that the act of observing probably has some disturbing effect.

Normal distribution. A frequency distribution of a "normal" group of things, measurements, dimensions. What does it take for a group to be normal? Hard to say, precisely, but fairly easy to point to an example: the heights of a large number of people (thousands) who have not been selected out for some special characteristic. It is easy to observe, in such a group, that there are few very tall persons in the group and few very short but more of each height as we approach the middle, where we find the largest numbers for each height. See "Frequency distribution."

Objective; objectivity. Objective observation would be such that other equally skilled observers would agree on the observations. Thus, one can be objective only about things that others might observe. Another person cannot observe my feelings nor can I observe his, so the feelings are subjective (as distinct from objective) or may only be experienced subjectively.

Placebo. In a blind or double-blind experiment, it may be necessary to lead a subject person to believe that he is getting a vitamin, medicine or other treatment when he is not. Instead he is given a nonacting substitute designed (it is hoped) so that he cannot distinguish it from the real thing. The substitute would be called a placebo.

Population. A large group from which a sample may be taken. In statistical work, "population" may refer to an unknown group which is defined only by reference to the sample in hand.

Predict; extrapolate. To make judgments beyond the actual observations or data at hand. To predict usually refers to a judgment about a future event. To extrapolate may mean to go beyond or, say, higher or lower than the observed range of the data.

Prejudice. An opinion, attitude or belief held in spite of evidence for a different view or with no real evidence that it has validity. A prejudice may be formed without any evidence or may be held even though contrary to the evidence.

Probability. A statement of the chance element in predicting a future event and, more accurately, a mathematical expression of the likelihood of a given thing happening. Probability may also be used to refer to past happenings as applied, for example, to the results of an experiment: Did the relation between the independent and dependent variables appear to be simply accidental (pure chance) or was there a "stronger" relation? One event may have a higher (or lower) probability of happening than another, or greater (or lesser) probability, or the event may be more (or less) probable.

Proof; prove. In this book I have avoided the use of the words "proof" or "prove" in discussion of the outcomes of investigations, or have enclosed those words in quotation marks. This I have done in recognition of the principle that the results of all investigations are "chancey" or are only more or less probable.

Random; randomize; random choice. One makes a random decision when the choice is made to depend upon the toss of a coin, the roll of dice, pulling names out of a hat, drawing a card out of a well-shuffled deck, using a table of random numbers, etc.

Range. In a series of measurements, or quantities, the difference or distance between the limits or extremes. For example, the range of ages in a group was from a 4-month-old baby to a 93-year-old grandmother. (Also see Figure G.1).

Replicate. To repeat an experiment or survey or other investigation; to make an investigation like the original to help to determine the validity of the original.

Sample. Used in statistical work with a meaning roughly equal to the common usage but also referring to highly developed methods for assuring that the sample is a fair representation of the population from which it was taken.

Scientific method. The use of observation of real things, events, people, to test hypotheses or to answer questions instead of using only reasoning or logic. Empirical testing or observation.

Statistics; statistical treatment. The rearrangement and analysis of data or factual material in an effort to get more meaning or information out of the data.

Subject (in an experiment). In an experiment, the person, animal or other thing changed or manipulated for the purpose of the test. Contrast with the experimenter who does the changing or manipulating.

Survey. An investigation of things as existing or of events past, to gain information or to test hypotheses, as contrasted with an experiment in which the things are changed or manipulated to observe the results.

Theory. An attempt to explain things or events with an explanation that has not been tested. Usually a theory is stated in rather general terms and is tested (if at all) by making up hypotheses related to it but specific enough to permit testing.

Variability. A statistical term expressing, for a group of values, the amount of variation around the mean of the values. Thus, a set of values may have the same mean and range as another set but greater or lesser variability. (See Figure G.1). The most widely used method for expressing variability mathematically is called "standard deviation."

Variable; related variables. In a thing (or pattern of things) being observed, a change in a dimension or a characteristic; a change in value or quantity. In an experiment, the experimenter makes a change (usually one but perhaps more) called the independent variable. He then observes any other variable that may develop. If this appears to be related to the independent variable or appears to depend upon it, he calls it a dependent variable and reports having found related variables.

Figure G.1
Mean, Range, Variability

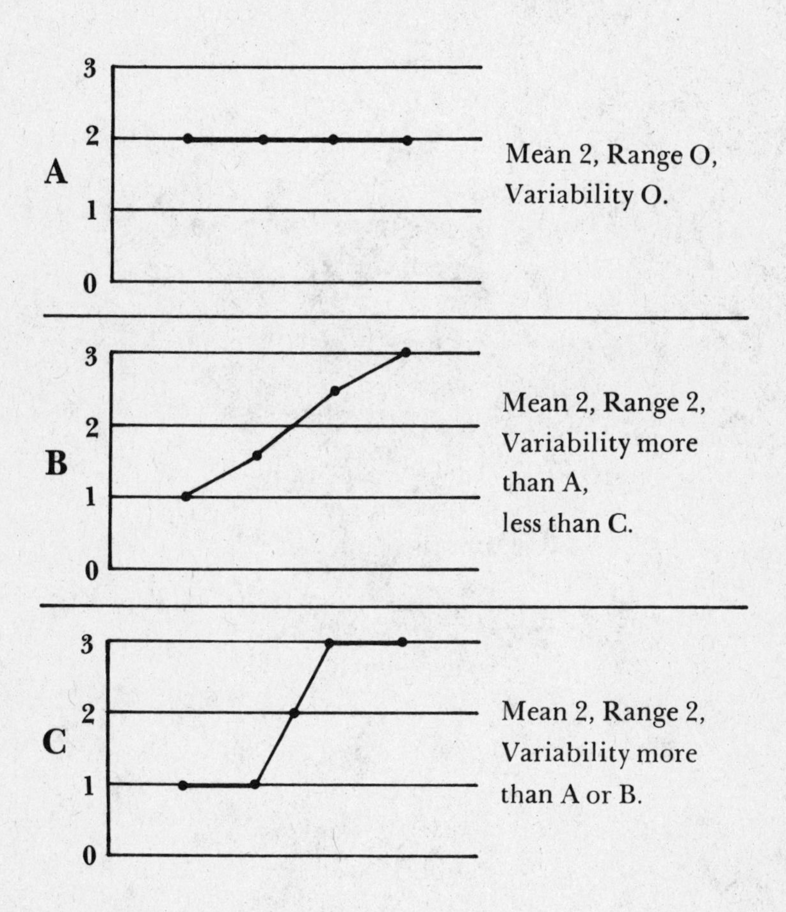

Compare A and B; they show that two groups of values may have the same mean but different ranges.

Compare B and C; they show that two groups may have the same mean and the same range but different variabilities.

Bibliography

There are hundreds of books published on experimental design, the design of surveys, and statistics for research. In this bibliography, I suggest a few that may be more readily available. These are divided, roughly, into a group which may be more useful to beginners and a group for those more advanced.

In the encyclopedias, look up articles headed: "Science," "Scientific Methods," "Experiment," "Survey," "Statistics," "Logic" and related articles.

For Beginners

Amos. J. R., Brown. S. L., and Mink, O. G., *Statistical Concepts:A Basic Program,* Harper & Row, New York, Paperbound, 1965, 124 pp.
Emphasizes concepts. A programmed text. An excellent review treatment for someone who has formerly studied statistics. Almost no mathematics, few applications. Not a manual.

Compton's Pictured Encyclopedia, Encyclopaedia Britannica, Chicago.

Collier's Encyclopedia, Crowell-Collier, New York.

Encyclopedia International, Grolier, New York.

World Book Encyclopedia, Field Enterprises, Chicago.

Youden, W. J., *Experimentation and Measurement,* Scholastic Book Services (for National Science Teachers Assoc.), New York, Paperbound, 1962, 127 pp.
For high-school people, or younger, showing problems in measurement and suggesting interesting experiments in measurement. Some statistical treatments described. Not a manual.

For the more advanced student or the professional

Cochran, W. G., and Cox, G. M., *Experimental Designs.* Wiley, New York, 1957.
An advanced professional work, widely quoted. Examples of research projects are worked out in detail with full statistical treatment and practical comments. Tables.

Diamond, S., *The World of Probability: Statistics in Science,* Basic Books, New York, 1964.

A discussion of statistics at high-school level. A popular treatment of concepts usual in a first course in statistics but not in enough detail to serve as a manual.

Encyclopedia Americana, Americana Corp., New York.

Encyclopaedia Britannica, Encyclopaedia Britannica, Chicago.

Fisher, R. A. *The Design of Experiments,* Hafner, New York, (First published 1935), 1966, 248 pp.

A classic in the field of experimental design and the logic of scientific inquiry. Largely verbal discussion; some mathematics. Not a manual; no tables.

Garrett, H. E., *Elementary Statistics.* David McKay. New York, 1962, 203 pp.

A beginning college text with emphasis on psychological and educational applications. Usable as a manual; includes tables.

McGraw-Hill Encyclopedia of Science and Technology, McGraw-Hill, New York.

Youden, W. J., *Statistical Methods for Chemists,* Wiley, 1951, 126 pp.

Examples discussed in a direct, agreeable style.

Index